THE A.I. ADVANTAGE

FOR AN INDIVIDUALIZED EDUCATION PROGRAM

A Parent's Playbook for Winning Results!

Al Jones, Jr., Ed.D.

leverage in **learning**

The A.I. Advantage for an Individualized Education Program: A Parent's Playbook for Winning Results
Copyright © 2025 by Al C. Jones, Jr.

Published by Leverage in Learning
http://www.leverageinlearning.com
Contact: info@leverageinlearning.com

Printed in the United States of America

For permissions, bulk orders, or bookings, please address correspondence to: info@leverageinlearning.com.

ISBN: 978-1-7354534-2-2

Dedication:

To the one who builds with logic,
loves with quiet strength,
and never wavers.

Table of Contents

Introduction

Welcome to the team, parents! If you're holding this book, it means you care deeply about a child's future—and you're navigating a process that can often feel overwhelming, confusing, and isolating. I understand.

As a lifelong educator, special education leader, and advocate—and as someone who has sat on both sides of the IEP table—I've seen firsthand how powerful, and how frustrating, this process can be. That's why I'm stepping in as your coach.

You need a winning game plan. That's what this guide is designed to help you create: a clear, focused, and adaptable strategy to meet your child's special education needs. Just like a great team doesn't step onto the field without a game plan, parents navigating special education shouldn't have to guess their way through it. You deserve a playbook that makes sense.

That's why I wrote this book.

You shouldn't have to be a lawyer, a teacher, and a therapist all in one to ensure your child gets a meaningful education. You just need the right tools, the right language, and someone to remind you that your voice matters. Think of me as your coach, guiding you step by step to build the strategies and confidence you need to advocate powerfully.

And it's not just about today. In the future, more and more responsibility for overseeing and carrying out special education law will shift to the local level as the federal footprint in education becomes smaller. That makes it even more critical that parents have a strong, working knowledge of how to advocate. You'll need a winning strategy that holds up, no matter how the system evolves.

This book is designed to be short, clear, and usable today. You can read it in one sitting—or one chapter at a time. It includes reflection questions, practical tips, and even the option to get customized help based on your answers.

And yes, you'll also learn how to use artificial intelligence—not to replace your insight, but to amplify it. Just like a great coach uses data and strategy to help their team win, you'll use A.I. to sharpen your advocacy skills and make sure your child's needs stay front and center. I believe technology, when used ethically and strategically, can level the playing field for families. It can help us communicate better, organize smarter, and advocate more clearly.

This guide is intended to assist you in devising a winning strategy for your child's special education needs. It's not just information—it's a handbook for action. My hope is that it strengthens your advocacy muscles and makes sure you never feel like you're alone in this. I hope it gives you confidence. And maybe even a little peace.

You're not alone. You're not powerless. And your child's IEP can—and should—reflect the brilliant, complicated, one-of-a-kind person they are.

So let's huddle up and get to work.

—Dr. Al C Jones, Jr.

CHAPTER 1

Understanding the IEP — A Parent's Superpower

E very great coach knows that success starts with preparation. As a parent, understanding the IEP process is your game plan for advocacy. This isn't just paperwork; it's your child's roadmap to success, and you're the one leading the charge.

Your child's IEP (Individualized Education Program) isn't just a document; it's a blueprint for success, a legal safeguard, and a tool for advocacy all rolled into one. But unless it's built well, it can be vague, underwhelming, or ineffective. That's why understanding how to shape your child's IEP is one of the most powerful things you can do as a parent.

An IEP is a legally binding document that outlines:

- Your child's current levels of performance

- Goals for growth over the next year

- The specialized instruction and supports they will receive

- How progress will be measured

- What accommodations and services are needed to succeed

It's more than paperwork; it's a contract between your child and the school, one that's backed by federal law.

By law, the IEP team includes:

- You (the parent or guardian)

- Your child (especially at transition ages)

- General and special education teachers

- A school district representative

- Specialists (speech therapist, psychologist, etc.)

> **COACHING PRO TIP**
>
> Your child's IEP team must include at least one general education teacher, even if your child spends most of the day in special education settings.

But you bring the most important voice: the lived experience of your child.

> **PARENT VOICE**
>
> **Tasha**
> "I remember sitting at that table, surrounded by professionals talking in acronyms and data I didn't understand. They were talking about my child, but I felt invisible. I left that meeting with a thick packet and a heavy heart. It wasn't until I learned how to read the IEP and ask questions that I started to feel powerful."
> — *Tasha, mother of a 3rd grader with ADHD*

Too often, IEPs fall short of what IDEA (the Individuals with Disabilities Education Act) intended. They are:

- Generic ("Johnny will improve reading comprehension.")

- Unmeasurable ("with 80% accuracy" but on what task?)

- Unrealistic or under ambitious

- Written with limited parent input or understanding

Although they recognize this about their child's IEP, many parents still walk into IEP meetings feeling powerless to do anything about it. But that stops now. I'm here to help you take charge, ask the right questions, and shape a plan that truly serves your child's needs.

You see your child in a way no teacher, evaluator, or administrator can. You know what frustrates them, motivates them, and makes them shine. When you're equipped with the right tools, you can drive the IEP process, not just respond to it.

This book is designed to guide you through that journey.

You'll learn how to collect meaningful information about your child, reflect on what's working and what's not, and, if you choose, submit your insights to a licensed special educator. That educator will create a fully customized, expert-guided IEP support plan you can take into your next meeting with confidence.

This is your moment to step into the leadership role your child needs. With the right strategy, you can make sure this process works for you — not against you.

COACH'S CORNER

Key Plays to Keep in Mind
- Remember that every strong IEP starts with a clear understanding of its core components.
- Accept that you are a vital member of the IEP team— your insights matter.
- Don't just view the IEP as just paperwork; view it as your child's legal right to support.

✅ ACTION STEP

Write It Out

Grab a notebook or open your notes app and list **every frustration** you've had with the IEP process—big or small. Name them, don't hold back.

Then, at the end of your list, write this out loud and proud:

"I am determined to overcome and win this challenge!"

Because you are not just a parent—you are your child's champion.

➡️ NEXT STEPS

- **Pull out your child's IEP** and keep it nearby as you read this book. You'll want to reference it often.
- **Make a list of questions** you have right now about your child's services, supports, or outcomes. Check them off as they're answered.
- **Still have questions at the end?** That's your sign to reach out. I'm here to help you go deeper and get the clarity you need.

✏️ QUICK TIP

Make This a Team Sport

Talk to others who care for your child—your spouse, a grandparent, an aunt, or a trusted friend. Ask them what they've noticed or what concerns they've had. Their observations might highlight strengths or needs you've missed.

Remember, great teams win together.

UNDERSTANDING THE IEP

A PARENT'S SUPERPOWER

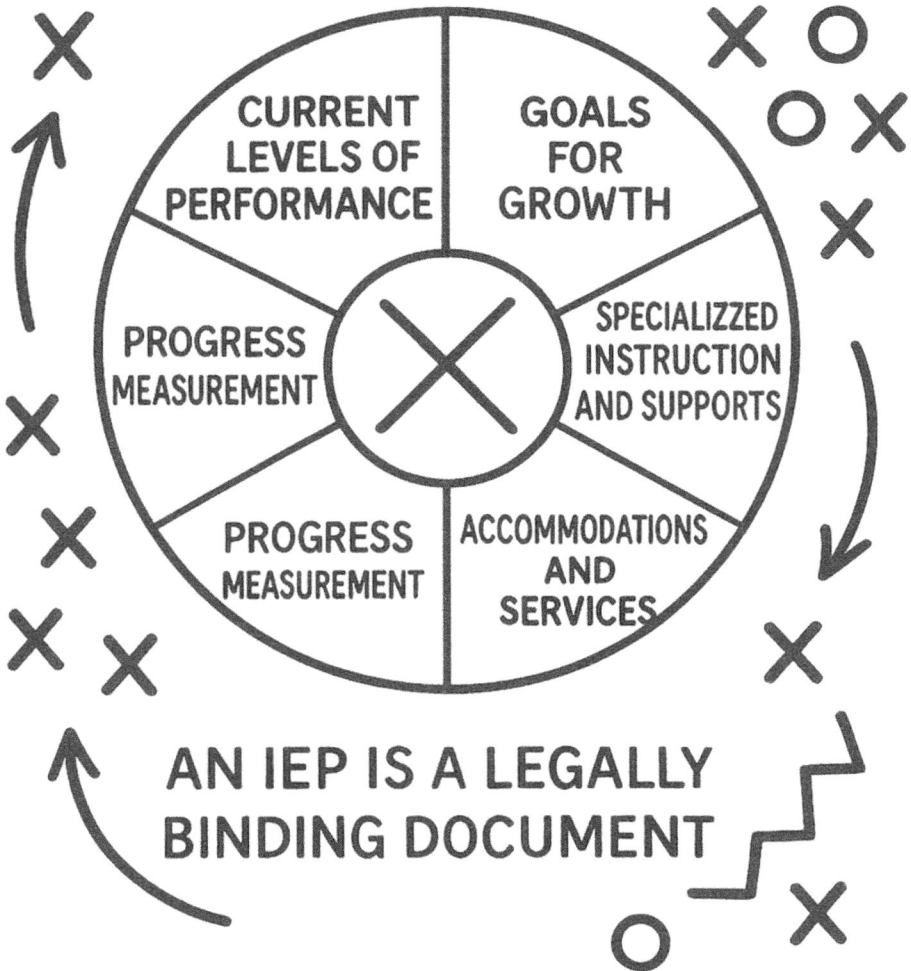

CURRENT LEVELS OF PERFORMANCE

GOALS FOR GROWTH

PROGRESS MEASUREMENT

SPECIALIZZED INSTRUCTION AND SUPPORTS

PROGRESS MEASUREMENT

ACCOMMODATIONS AND SERVICES

AN IEP IS A LEGALLY BINDING DOCUMENT

CHAPTER 2

Why Most IEPs
Fall Short — and
How to Fix Them

Every team needs a playbook that details their strategy. As your child's biggest advocate, you need one, too. If you've ever left an IEP meeting feeling lost, I get it. But today, I'm going to help you change the game.

If you've ever walked out of an IEP meeting feeling unsure, frustrated, or overwhelmed, you're not alone.

Although legally required to support a child's unique needs, many IEPs fall short. They're often written quickly, recycled from previous years, or filled with generic language that doesn't reflect your child's strengths or challenges.

But here's the good news: once you understand *why* IEPs fall short, you'll be better equipped to build one that truly works.

TINY STORY

Hidden Potential:

When Marcus's mom read through his IEP, she noticed all the goals were labeled 'basic.' But at home, Marcus was using a keyboard to write full sentences. Once she brought in video clips and work samples, the team revised his goals—and Marcus's progress exploded.

The 5 Most Common Ways IEPs Fall Short

1. Vague or Generic Goals

DID YOU KNOW?
Many schools use template-based IEPs to save time—but this can result in generic language that doesn't reflect your child's individual strengths and needs.

Too often, IEP goals are written in broad terms that don't provide a clear path forward. A weak game plan leads to weak results. You wouldn't train an athlete without tracking progress, so why settle for vague goals?

For example, an IEP goal might state: "The student will improve reading comprehension with 80% accuracy."

What does that mean? What task? What level of text? How is it measured?

Fix it: Insist on goals that are *specific, measurable, achievable, relevant*, and *time-bound* (SMART). Each goal should include the task, baseline data, and clear criteria for mastery.

Try this: Pick one of your child's current goals and rewrite it using SMART criteria.

Example: The Vague Goal Trap
"One IEP said Marcus would 'improve his reading skills.' When his dad asked, 'How will we know when he's improved?' the team revised it to: 'Marcus will read a third-grade passage with 90% accuracy across three sessions."

Example: Using Strengths in the IEP

"Sasha's IEP described her as 'struggling in math,' but her mom shared that Sasha loved baking and often doubled recipes on her own. That detail helped the team write a math goal tied to real-life application."

2. Lack of Alignment to Grade-Level Standards

QUICK TIP

Ask your IEP team: "Can you show me how this goal aligns with grade-level standards?" This keeps your child on track with peers and ensures high expectations.

Fix it: Sometimes goals are written far below grade level, which limits expectations and progress.

Request goals that align with state standards, even with scaffolding if necessary.

3. Limited Input from Parents

Many schools come to the meeting with a draft already written — and your voice may be an afterthought.

Fix it: Prepare in advance. Bring your own observations, examples, and concerns. Your input is not optional — it's essential.

PARENT VOICE

Goals

I didn't realize I could ask for the data behind my child's goals. Once I did, the team revised them to actually fit where she is—not where they assumed she'd be.

—Parent of a 2nd grader with a language-based learning disability

4. No Connection Between Services and Goals

It's not uncommon for the services listed in the IEP to seem unrelated to the goals.

Fix it: Ask directly: "How does this service support this goal?" Each service should be purposeful and clearly connected to your child's specific learning objectives.

PARENT VOICE

Services

"They told me we'd have to wait until next year to change the services. But I found out I could request a meeting at any time—and we got more support within weeks."
—Single parent of a child with autism

5. Missing Progress Monitoring

QUICK TIP

Ask: "What tools will be used to track progress, and how often will we review the data?" Progress reports should be as frequent as report cards.

Fix it: Without regular, clear data, there's no way to know if your child is making meaningful progress.

Request frequent, data-based progress reports. Ask what tools will be used, how often progress will be tracked, and how it will be shared with you.

> **PARENT VOICE**
>
> **Data**
>
> *I didn't realize I could ask for the data behind my child's goals. Once I did, the team revised them to actually fit where she is—not where they assumed she'd be."*
>
> —Parent of a 2nd grader with a language-based learning disability

How to Take Back Control: Your 3-Step Fix Framework

You don't have to be a special education expert to contribute powerfully to your child's IEP. You just need a system that helps you show up prepared and focused.

Step 1: Know Your Child's Story

Use the questionnaire included in this book to collect key observations and insights. These paint a fuller picture of who your child is and what they need.

Step 2: Submit Your Input Before the Meeting

Don't wait for the meeting to share your thoughts. Send your insights ahead of time. You can also use our service to create a polished draft tailored to your child's needs.

Step 3: Ask Better Questions in the Meeting

A powerful question shifts the tone of the meeting. Try:

- "How will this help my child meet this goal?"
- "What will progress look like?"
- "How often will we review it together?"

These questions move the team beyond compliance and toward real collaboration.

Final Word: You're Not Just a Participant... You're the *Leader*!

You are the one constant in your child's educational journey. You know your child best.

Come informed.

Be confident.

Lead with purpose.

COACH'S CORNER

Key Plays to Keep in Mind
- A SMART goal is a strong goal—vagueness is your enemy.
- Push for alignment between goals, services, and grade-level standards.
- Don't wait to be invited—lead with your observations and your voice.

NEXT STEPS

Review your child's current IEP with fresh eyes. Highlight what's clear, what's vague, and where you want change.

Remember, this isn't just about fixing an IEP. It's about coaching yourself into a confident advocate. Now, let's get ready for the next play!

📖 What this chapter helped you do

This chapter equipped you with the tools to spot weaknesses in your child's IEP and take actionable steps to fix them. You now know how to evaluate goals, request alignment with standards, advocate for meaningful services, and demand clear progress monitoring. You're not just reacting—you're leading with strategy and confidence.

MOST IEPs DO NOT MEET STUDENTS' NEEDS BECAUSE THEY:

① lack individualized goals

② overlook key skill areas

③ provide ineffective instruction

④ lack meaningful progress monitoring

CHAPTER 3

What Schools Won't Tell You — But You Need to Know

I n any strong game plan, knowing the rules changes everything. When it comes to your child's education, understanding the system is your advantage.

Even the most well-meaning educators don't always share the full picture when it comes to IEPs. Sometimes it's because they assume you already know. Other times, it's because they don't know themselves. And occasionally, it's because of internal pressures, like limited time, training, or resources.

But as a parent, information is power. The more you understand how the system works, the better you can navigate it, and the more effectively you can advocate for your child.

Here's What You Need to Know

1. Not All IEPs Are Created Equal

IEPs can vary dramatically from school to school, even within the same district. Some schools have experienced teams and strong systems. Others are short-staffed or may lack familiarity with your child's specific disability.

ACTION STEP

Ask, *"Has the team worked with a child with this profile before?"* If not, don't hesitate to request outside evaluations or support. You're allowed to ask for what your child needs, even if the school isn't already offering it.

> **PARENT VOICE**
>
> *IEPs*
>
> *"We'd been using the same IEP template for three years. It wasn't until I asked them to walk me through each goal that we saw how little had changed. That meeting changed everything."*
>
> —Father of a 6th grade student with ADHD

2. You Have More Power Than You Realize!

Federal law under IDEA gives you powerful rights, including the ability to:

- Participate in every decision about your child's education

- Request evaluations

- Call an IEP meeting at any time

- Disagree with decisions and request mediation or due process

DID YOU KNOW?
You can request an IEP meeting at any time—you don't have to wait for the annual review. Just submit the request in writing.

These rights exist to ensure your voice matters — not just in theory, but in practice.

ACTION STEP

If something doesn't feel right, say something. You don't need permission to advocate. You already have the right.

Try this: Write down three rights under IDEA that you didn't fully know before reading this chapter. Which ones increase your confidence or give you new clarity on how to support your child?

3. IEPs Must Be Based on Data — Not Just Opinion

A strong IEP is built on measurable data, not guesswork. That includes test scores, observation reports, work samples, and consistent progress monitoring.

ACTION STEP

Ask, *"What data was used to make this recommendation?"* If you hear general statements instead of specifics, keep asking until you get clarity.

4. Budget Constraints Are Not an Excuse

DID YOU KNOW?
It's illegal for a school to deny needed services based on cost alone. FAPE (Free Appropriate Public Education) is not optional—it's the law.

Schools are legally required to provide a Free Appropriate Public Education (FAPE) even when resources are tight. Cost should never be the reason your child isn't getting what they need.

ACTION STEP

If a service is denied due to cost or staffing, ask for that denial in writing. This request can trigger a formal review, and it often changes the outcome.

Fix It: If the school says they "don't have the funds" for a support your child needs, say this:

"I understand budget constraints are real. But I believe this service is necessary for my child to receive FAPE. Can you please put the reason for this denial in writing so I can consider my next steps?"

TINY STORY

Learning the Law, Changing the Outcome

When Monique's daughter was denied additional reading support due to staffing limits, she didn't argue in the moment. Instead, she went home, read up on IDEA, and wrote a letter quoting the law's requirements. The school responded within days, offering a compromise plan that included extended learning time. Her voice—and knowledge—made the difference.

Example: Budget Constraints

"When Alina's school said they didn't have the resources for additional speech therapy, her grandfather calmly responded, 'I understand the challenge—but federal law requires FAPE regardless of cost. Can we look at other funding options?' The team reconvened a week later with a solution."

5. You're Allowed to Bring Help

You don't have to face an IEP meeting alone. You can bring anyone you trust: a spouse, a friend, a therapist, an advocate, or even a professional consultant.

✅ ACTION STEP

If you're feeling overwhelmed, bring someone. Another set of ears can make all the difference… and having support may give you the confidence to speak more freely.

✏️ QUICK TIP

If a meeting feels rushed or dismissive, pause and redirect. Say something like:
- "I'd like to take more time to process this. Can we reconvene to give these concerns the attention they deserve?"

Or, if you feel unheard:
- "I'm feeling like my input isn't being fully considered. I'll be following up with a written summary and cc'ing the principal to ensure our discussion is documented."

COACH'S CORNER

Key Plays to Keep in Mind
- Knowing your rights changes how you show up—own your role as a decision-maker.
- FAPE is not optional—budget concerns cannot override your child's legal protections.
- Written communication protects your advocacy—put concerns and requests in writing.

The Bottom Line: You're Not a Guest, You're the *Leader*!

The more you understand your rights, the more confidently you can walk into your next IEP meeting. You're not asking for favors. You're functioning in your role, not as a passive participant, but as a leader in your child's educational journey.

- Most IEPs fail because they lack specificity, alignment, or measurable outcomes.

- You can spot a weak IEP by looking for vague goals and missing accommodations.

- You have the power to advocate for clarity, rigor, and relevance.

What Schools May Not Tell You—but You Should Know:

- You can call an IEP meeting at any time—not just annually.

- Budget is not a legal excuse to deny services.

- Your child's IEP should evolve as they grow and learn

TINY STORY

Asking the Right Questions:

Sofia's dad didn't know what to ask—he just knew things weren't working. After using the questionnaire, he realized no one had ever documented how noise impacted Sofia's reading. They added sensory accommodations and saw a change within weeks.

Great teams win because they prepare. You've got the knowledge. Now, it's time to step into the meeting with confidence!

What this chapter helped you do

This chapter increased your awareness of your legal rights under IDEA and helped you see how to respond strategically when schools fall short—especially when citing budget or limited resources. You've learned how to request meetings, ask the right questions, bring in help, and hold the system accountable. You now have the language and mindset to lead your child's IEP process with purpose and clarity.

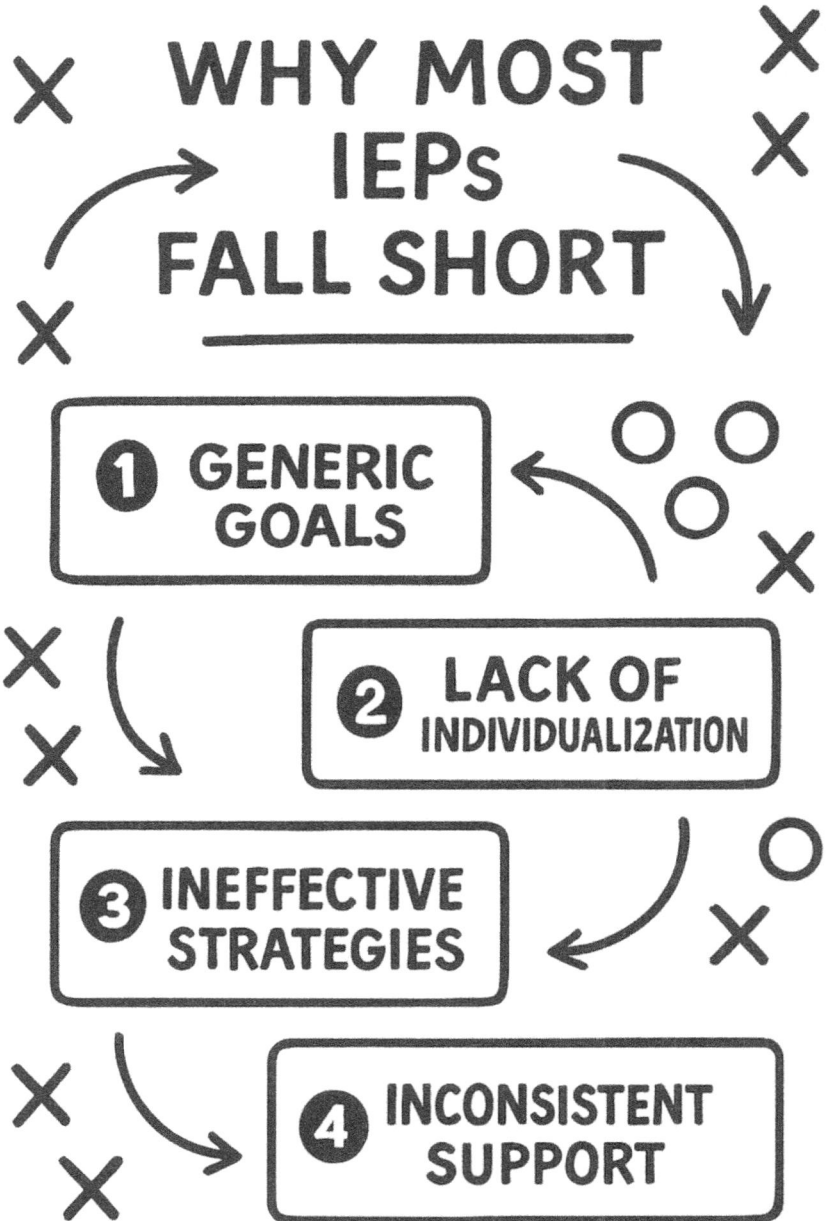

WHY MOST IEPs FALL SHORT

1 GENERIC GOALS

2 LACK OF INDIVIDUALIZATION

3 INEFFECTIVE STRATEGIES

4 INCONSISTENT SUPPORT

CHAPTER 4

The 40 Questions That Will Transform Your Child's Education

Y ou've probably been to IEP meetings where it felt like you were "filling in the blanks" on a document you didn't create.

That changes today.

This chapter gives you a structured questionnaire designed to help you reflect on what's working, what's not, and where your child needs more personalized support. These 40 questions form the foundation of any strong IEP.

Great teams win with preparation. If you've ever felt like an observer in IEP meetings, these questions will shift you into the driver's seat. They can guide team meetings, improve communication with your child's school, and support planning at home.

DID YOU KNOW?
Federal law requires that every IEP include measurable goals, services, and how progress will be reported—but studies show more than 60% of IEPs reviewed in audits lack clear progress measures.

TINY STORY

The Power of a Questionnaire

"That questionnaire helped me organize everything I'd been trying to say for years. I felt like I finally had a roadmap instead of a mess of notes and emails."

—Parent advocate and foster parent

How to Use This Chapter

- Reflect and respond on your own, or with a partner.

- Use these questions before your annual IEP meeting or whenever new concerns arise.

- You can also submit them through Leverage in Learning to receive a custom IEP improvement plan based on your responses.

TINY STORY

A Structured Approach Changes Everything

"Darren's mom always left IEP meetings frustrated—until she came in with three specific questions from this list. When she asked, 'How will this goal be measured?' and 'Can you show me the data used to set this level?' the tone of the meeting shifted. For the first time, she saw the team treating her as an equal. And for the first time, the IEP actually made sense."

Fix it: Refine Before You Submit

Once you've answered the questions, review your responses and ask:

- Is this clear to someone who doesn't know my child?

- Did I give specific examples or just general feelings?

- Would this response help someone write a goal or make a decision?

If your answer is vague (e.g., "Math is hard"), try revising it to something more actionable (e.g., "My child struggles with multi-step word problems and often leaves math worksheets unfinished due to anxiety.")

The more detailed and accurate your answers, the more effective your IEP support will be—whether you bring it to the meeting yourself or submit it for professional review.

> **PARENT VOICE**
>
> **Questionnaire**
> "That questionnaire helped me organize everything I'd been trying to say for years. I felt like I finally had a roadmap instead of a mess of notes and emails."
> —Parent advocate and foster parent

40 Questions

Section 1: Your Child's Profile

1. What is your child's full name and date of birth?

2. What grade is your child currently in?

3. What school and district do they attend?

4. What is your child's current disability classification (if known)?

5. What do you want the IEP team to understand most about your child?

Section 2: Strengths and Interests

6. What are your child's academic strengths?

7. What are their personal or social strengths?

8. What do they love to do outside of school?

9. What motivates them to try hard or stay focused?

10. How does your child learn best (visuals, repetition, hands-on, etc.)?

Section 3: Academic and Functional Needs

11. What academic skills are the most challenging right now?

12. How does your child handle frustration or confusion in school?

13. Are there subjects where your child is consistently behind grade level?

14. Are there skills your child seems to be "stuck" on year after year?

15. Are there functional (non-academic) challenges such as organization, social interaction, self-care, or behavior?

Section 4: Current IEP Goals and Services

16. Does the current IEP include goals that match your child's needs?

17. Are the goals written clearly and in a way that makes sense to you?

18. Do you think the goals are challenging enough—or too easy?

19. Do the current services (e.g., speech, OT, counseling) appear to help?

20. Are you satisfied with the frequency and quality of services being delivered?

Section 5: Inclusion and Access

21. Is your child included in general education classes when appropriate?

22. Does your child have the support they need to be successful in those classes?

23. Do you believe your child is being given access to the same curriculum as peers?

24. Are there social opportunities your child is missing out on?

25. Does your child have friendships or positive peer relationships at school?

Section 6: Communication and Collaboration

26. How often do you receive updates about your child's progress?

27. Are you satisfied with the communication you receive from the school?

28. Do you feel like a full and equal member of the IEP team?

29. Have your concerns been taken seriously in past IEP meetings?

30. Is there a staff member at school you feel comfortable talking to?

Section 7: Home-Based Learning and Support

31. What strategies or routines work well at home?

32. What academic skills do you help your child with at home?

33. Does your child complete homework with independence or support?

34. How do you motivate your child to stay focused at home?

35. Are there areas where home and school could coordinate better?

Section 8: Vision and Long-Term Planning

36. What do you hope your child will achieve by the end of this year?

37. What do you envision for your child 3–5 years from now?

38. What's your biggest fear or concern about your child's future?

39. What's your biggest hope for your child's future?

40. If you could change one thing about the current IEP, what would it be?

QUICK TIP

Before your next IEP meeting or check-in with the school, pick **three to five questions** from this list and bring them with you. Use them to guide the conversation, raise concerns, or clarify what's missing. Questions shift the tone of meetings from passive to purposeful.

COACH'S CORNER

Key Plays to Keep in Mind
- Reflect deeply on your child's needs and your goals as a parent.
- Gather 40 responses that can shape a stronger, more focused IEP.
- Get ready to bring data and insight into your next meeting.

NEXT STEPS

- Review and complete the 40-question parent questionnaire
- Highlight the 3–5 priorities you want to bring to your next IEP meeting
- Decide whether to proceed on your own or submit your responses to Leverage in Learning for help
- Use the tips, templates, and prompts to prepare your next conversation with your school team

These Questions Are Your Advocacy Power

Whether you complete this list on your own or share it with your child's team or a professional advocate, you've just created a comprehensive map of your child's unique needs, strengths, and story. This is the raw material for a truly individualized IEP.

If you'd like help turning this into a personalized plan, submit your completed responses to acjonesjr@leverageinlearning.com. Together, we'll build a smarter, stronger plan that reflects your child, not just their paperwork.

This questionnaire isn't just a checklist; it's your strategy for success. Now, let's get you ready to lead the conversation!

TINY STORY

More Than Just a Form:

Lena filled out the questionnaire during her lunch breaks. It was tedious—but when she got the final plan back, she cried. For the first time, someone had seen her daughter's strengths and not just the disability.

IEP QUESTIONS FOR PARENTS

(40)

KEY QUESTIONS TO INFORM CUSTOMIZED IEP FOR YOUR CHILD

- ✓ STRENGTHS
- ✓ NEEDS
- ✓ SIGNIFICANT CONCERNS
- ✓ EFFECTIVE STRATEGIES

CHAPTER 5

Submitting Your Questionnaire — What Happens Next?

I f you've completed the parent questionnaire, congratulations! You've just built the foundation for a game-changing IEP that truly reflects your child's needs, strengths, and future potential.

Whether you decide to take the next steps on your own or use an expert support service like Leverage in Learning, here's what comes next.

Step 1: Review and Organize Your Responses

Take a moment to look back at your answers. Do they clearly describe your child's abilities, needs, and learning environment? Are there areas where you want to add detail or clarify your thoughts?

> **COACHING PRO TIP**
>
> Highlight 3–5 key priorities you want to focus on in your next IEP meeting. These might include an academic goal, a needed service, or an inclusion opportunity.

Step 2: Choose Your Path

You've got two strong options – both valid, both effective. The difference comes down to how much time and support you need.

Option 1: The DIY Approach

Use your completed questionnaire as a tool to guide the next IEP meeting.

- Share your responses with the IEP team ahead of time

- Highlight your child's strengths and suggest 1–2 specific goals

- Ask the team how services can support those goals directly

- Use your answers to shape the language and priorities of the IEP

This option is perfect for families who feel confident leading the conversation and want to maintain full control of the process.

Option 2: Get Expert Support with Leverage in Learning

If you'd like help transforming your insights into a polished, expert-designed IEP improvement plan, Leverage in Learning is here for you.

We offer personalized services tailored to your child's unique profile, whether you're preparing for your first IEP meeting or fine-tuning a plan that's already in place.

Visit www.leverageinlearning.com to:

- Explore our IEP plan-building options

- Choose the level of support that fits your needs

- Access tools and tips to use at home and in meetings

You can also upload your completed questionnaire directly, and we'll take it from there.

All plans are created by Dr. Al C. Jones, Jr., a national expert in special education and AI-integrated learning. Each plan is designed to align with federal IEP standards and the best practices in inclusive education.

Whether you're looking for a second opinion or want to walk into your next meeting fully prepared, we're here to help.

Step 3: Bring It to the Table

Once you've got your plan, whether self-made or professionally designed, bring it into the IEP conversation with confidence. This is your playbook, and you're stepping in as the lead strategist. Walk in prepared, ask the right questions, and shape the outcome.

- Print copies for yourself and the full IEP team

- Practice how you'll introduce your key priorities

- Ask, "Can this language be included in the final IEP?"

Fix it: If the school resists incorporating your priorities, stay calm and clear. Say:

"These priorities are based on observation, current needs, and IDEA guidelines. If there's a reason these cannot be added, can you please explain that reasoning and include it in the meeting notes?"

Asking for an explanation in writing helps ensure transparency and can prompt more thoughtful collaboration.

QUICK TIP

Before your next meeting, rehearse how you'll introduce your top 3 priorities in one or two clear sentences each. Rehearsing aloud builds confidence and helps you stay grounded, even if the conversation becomes tense or emotional.

✅ ACTION STEP

What to Do After the Meeting

- **Write it down while it's fresh.** In the car or right when you get home, jot down key takeaways, follow-up tasks, and anything left unresolved.

- **Send a follow-up email.** Thank the team for their time and recap the priorities you discussed. Confirm any next steps agreed upon.

- **Track your child's progress.** Over the next few weeks, watch for changes in services, communication, or outcomes related to what was discussed.

- **Plan your next move.** If your concerns weren't addressed, consider requesting a follow-up meeting—or reaching out for expert guidance.

Final Word: Your Voice Matters

This process isn't just paperwork, it's a declaration of what you believe your child deserves.

You've done the reflection. You've gathered the insight. Now, you're ready to take the lead.

No matter which path you choose, the key is this:

Be prepared. Be clear. Most importantly, remember this: you are your child's most important advocate.

COACH'S CORNER

Key Plays to Keep in Mind

- A plan on paper means nothing until you bring it into the room—lead the conversation.
- Rehearsing your top priorities helps you stay calm and clear under pressure.
- Follow-up is part of advocacy—clarity after the meeting is just as important as confidence during it.

NEXT STEPS

- Review your answers and highlight the top 3–5 priorities.
- Choose whether to use the DIY route or get expert help.
- Use your responses to lead your next IEP meeting with clarity.

You've done the work. Now, step into the meeting with the confidence of a coach rallying the team.

TINY STORY

Big Wins at Home:

Jamal struggled with math at school, but his aunt started using flashcard games at home based on his IEP goals. His confidence grew, and by spring, he was answering questions in class for the first time.

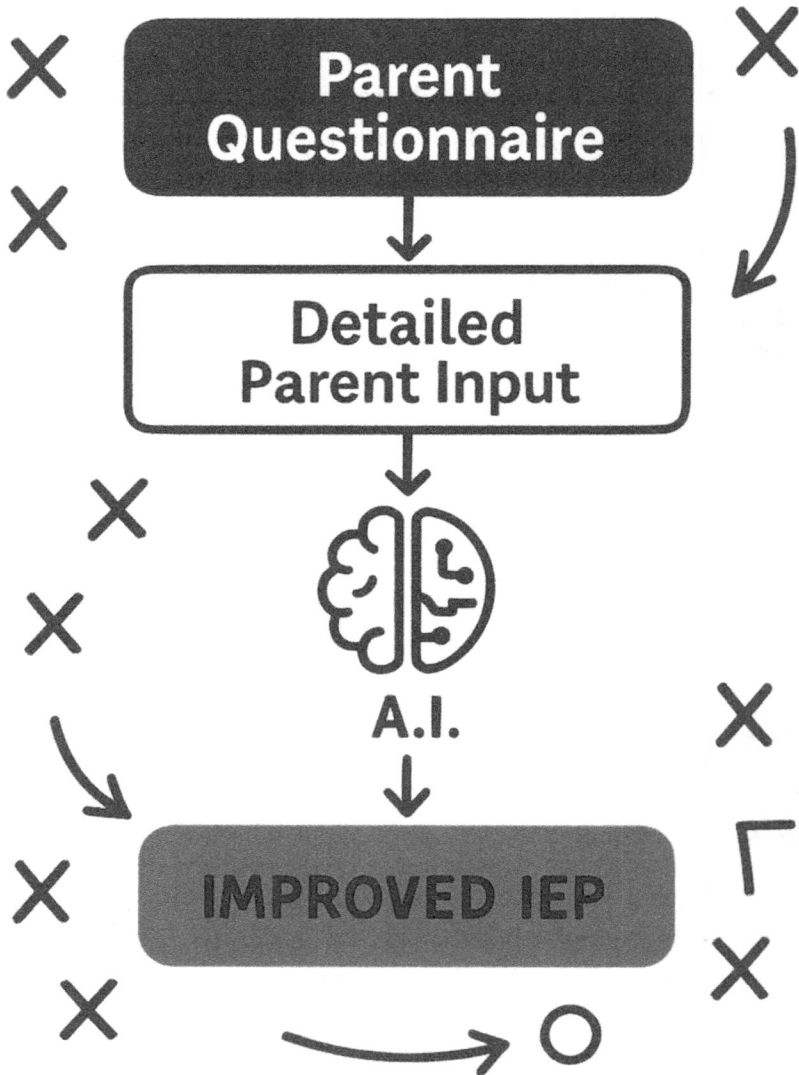

CHAPTER 6

The Power of Home Support — What You Do Matters

IEPs don't end when the school bell rings.

In fact, some of the most powerful growth happens after your child gets home.

When schools and families work together intentionally, consistently, and creatively, children with disabilities don't just improve; they thrive. And here's something many parents need to hear:

You have more influence than you think.

You're the MVP in your child's learning journey, so what you do makes a bigger difference than you realize.

You don't need to be a teacher. You don't need a curriculum.

You just need to know that what you do – even in small doses – makes a real difference.

Think of home as your training ground. Just a few minutes of intentional effort each day can transform how your child grows and learns.

Here's Why Your Involvement Matters

1. Reinforcement is Key

✏️ **QUICK TIP**

If your child is working on a skill at school, try reinforcing it at home for just 10 minutes a day. That small time investment can double retention.

Children with disabilities often need repetition across environments to truly master a skill. Learning something once a week at school isn't enough—but practicing it at home, even briefly, can lead to lasting growth.

QUICK TIP

If your child is working on comprehension, ask:
"Can you tell me three things that happened today?"
This builds sequencing, memory, and conversation skills—naturally and without extra pressure.

2. Home Is a Low-Stress Learning Environment

School can be overwhelming: noisy, fast-paced, and full of transitions.

Home is different. It's predictable. It's emotionally safe.

That's why your child may be more willing to try hard things, take risks, and bounce back from mistakes at home, especially when you're encouraging them.

QUICK TIP

Celebrate effort, not just results.
Saying "I love how you stuck with that even when it was hard" teaches grit and self-belief.

3. You See What Others Don't

You know your child in ways no teacher, therapist, or evaluator ever could.

You notice how they learn in the car, during dinner, or before bed.

You see the patterns across days, moods, and settings.

This perspective is crucial—because it tells the IEP team what's *really* working and what's not transferring beyond the classroom.

QUICK TIP

Keep a short "learning log."
Even just 2–3 notes per week (like "He asked for help today without melting down!") can guide smarter supports.

4. Your Voice Shapes the IEP

When you share what's happening at home, the school team can create better, more personalized goals and services. This isn't overstepping—it's partnership.

QUICK TIP

Bring examples of home learning to the IEP meeting: Photos, routines, a favorite book, or even a note your child wrote can spark meaningful conversation and collaboration.

Final Word: Progress Doesn't Only Happen at School

Supporting your child at home isn't about doing everything.

It's about doing small things that matter, consistently and intentionally.

And it's okay to ask: *"Which small things will matter most for my child?"*

That's what your IEP team and services like Leverage in Learning are here to help with on your journey to support your child.

You don't have to do this alone.

But know this: what you do matters. More than you know.

Try this: Pick one of your child's current goals and rewrite it using SMART criteria.

Struggling with Home Learning Consistency?

If your child struggles to focus or follow through with learning at home, try using a simple visual schedule or checklist.

Start small—pick one time of day (like after snack or right before dinner) and assign just one learning task (e.g., "Practice reading for 5 minutes").

Add a visual cue (like a timer or sticker chart) and end with something fun or relaxing to reinforce the habit.

This builds both structure and motivation—two powerful ingredients for learning at home.

TINY STORY

When AI Clicked:

After school, Noah's sister started using an AI chatbot to practice his math goals in a fun quiz format. It felt like a game—but aligned exactly with his IEP objectives. His teacher noticed the improvement in just two weeks.

NEXT STEPS

Choose **one small strategy** from this chapter to try at home this week (e.g., start a learning log, use a visual timer, or set a daily reading goal).

- Write down one success story or challenge from that experience.
- Bring that insight into your next IEP meeting—it can help shape goals and services that match your real life.
- If it helps, email your child's teacher to let them know what you're trying and what you're noticing.

COACH'S CORNER

Repetition at home doesn't need to be long—it just needs to be consistent.

- Learning can happen during everyday routines: in the car, at dinner, during chores.
- Your home experience is data—bring it to the IEP table with pride.

What this chapter helped you do

This chapter empowered you with practical strategies to support your child's learning at home and built your confidence to share that insight with your IEP team. You learned how small, consistent actions at home can strengthen your child's skills—and how your observations and routines can shape stronger, more personalized IEPs.

You've got the tools and the vision. Now, let's make every moment count!

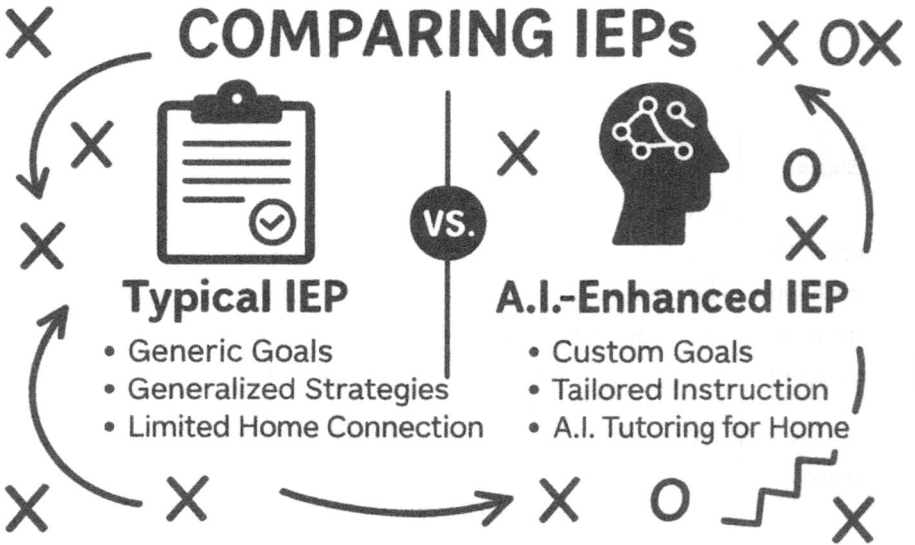

COMPARING IEPs

vs.

Typical IEP

- Generic Goals
- Generalized Strategies
- Limited Home Connection

A.I.-Enhanced IEP

- Custom Goals
- Tailored Instruction
- A.I. Tutoring for Home

CHAPTER 7

Using AI to Support Your Child's IEP at Home

L et's talk about one of the most powerful tools available to families today — and one of the most underused in special education: Artificial Intelligence (AI).

No, not robots teaching your child.

We're talking about accessible, user-friendly tools (like ChatGPT or speech-based assistants) that can support, tutor, and encourage your child right at home.

When guided by your child's IEP goals, AI can become a helpful learning ally, one that's available 24/7, doesn't get tired, and never judges.

Think of AI as your assistant coach: it's always ready, always adaptable, and here to help your child win at learning.

How AI Can Support Learning at Home

1. Practice Through Prompts

DID YOU KNOW?
AI tools like ChatGPT can simulate one-on-one tutoring, practice IEP goals, or even role-play social situations—just by typing the right prompt.

With just a few words, AI can become a friendly, patient tutor, asking questions, offering encouragement, and helping your child stay engaged.

📖 Example: Prompt for ChatGPT

"Act as a friendly tutor. Ask my child three reading comprehension questions about this short story. Help them if they get stuck."

Why it Works: AI responds in real-time, adapts to your child's answers, and keeps the tone positive. No eye-rolls, no pressure.

2. Repetition Without Resistance

Let's be honest: kids don't always want help from a parent.

But give them a tablet or voice assistant? Suddenly, they're *all ears*.

AI turns practice into play by making learning interactive, fresh, and fun.

✏️ QUICK TIP

Create a "tutoring corner" at home.
Call it something cool like the "Brain Boost Zone" and keep sessions short (5–10 minutes). Consistency matters more than duration.

3. Guided Practice with You or a Sibling

You don't need to be an expert. With a few simple scripts or help from an AI, you can guide short learning sessions at home.

📖 Example:

"Today we're working on using a number line to subtract. Let's try 10 – 4. Draw it out and walk me through your steps."

Older siblings can help too. With a script in hand, they become part of your child's learning team.

4. Customized Practice Built from Your IEP

When you use Leverage in Learning's custom IEP planning service, you'll receive AI tutoring prompts tailored to your child's specific goals.

These aren't generic worksheets; they're personalized scripts and tools based on your questionnaire, designed to meet your child *exactly where they are.*

Prompts grow with your child. As skills improve, we adapt the so they keep building confidence and independence.

Fix it: What if AI Prompts Don't Work at First?

If your child loses interest or the AI seems "off," don't give up—adjust the prompt. Try:

- Making the tone more fun ("Pretend you're a silly pirate teaching math!")
- Shortening the task ("Ask just one question about the story")
- Adding support ("If my child doesn't respond, give a hint.")

Prompt writing is like coaching—tweak it based on what your child responds to. And if all else fails, let them help design the next prompt. Ownership boosts engagement.

PARENT VOICE

AI in Action

"Using the prompts with my daughter at home gave me so much more confidence walking into her IEP meeting. I had proof she could do more than the school expected."

—Mom of a 4th grader with dyslexia

AI Isn't Replacing You; It's Empowering You!

AI doesn't take the place of a parent's care, wisdom, or insight.

But it *can* become another powerful tool in your parenting toolbox.

Used wisely, AI helps:

- Reinforce IEP goals

- Make learning consistent across school and home

- Reduce stress by giving you a guided plan

And most of all, it puts you in the driver's seat.

You don't need to know everything.

You just need to know this: you have tools, you have support, and you've got this!

You're building an unstoppable advocacy strategy. AI is just one tool, but when used right, it can reinforce your playbook for success.

📖 TINY STORY

A Turnaround Meeting

Dana felt like every IEP meeting was a battle. But when she brought in a written parent statement and her own copy of the goals, the team shifted. Instead of pushing back, they started listening—and collaborating.

NEXT STEPS

Try This This Week

- Choose one IEP goal your child is working on.
- Create a simple AI prompt to practice that goal (e.g., "Act as a math coach and quiz my child on subtraction facts under 20").
- Try the prompt with your child for 5–10 minutes.
- Jot down what worked and what didn't—this data helps shape future learning sessions *and* your next IEP conversation.

COACH'S CORNER

Key Plays to Keep in Mind
- Start small. AI is most powerful when it supports one focused goal at a time.
- Prompt quality matters—adjust wording based on your child's interest and attention.
- You don't need to know all the tech. You just need to try—and adapt as you go.

What this chapter helped you do

This chapter empowered you with AI strategies to support your child's IEP goals at home. You now understand how to use AI prompts to reinforce learning, how to troubleshoot when things don't go smoothly, and how to turn small, consistent tech-based routines into real educational gains.

COLLABORATING WITH THE IEP TEAM

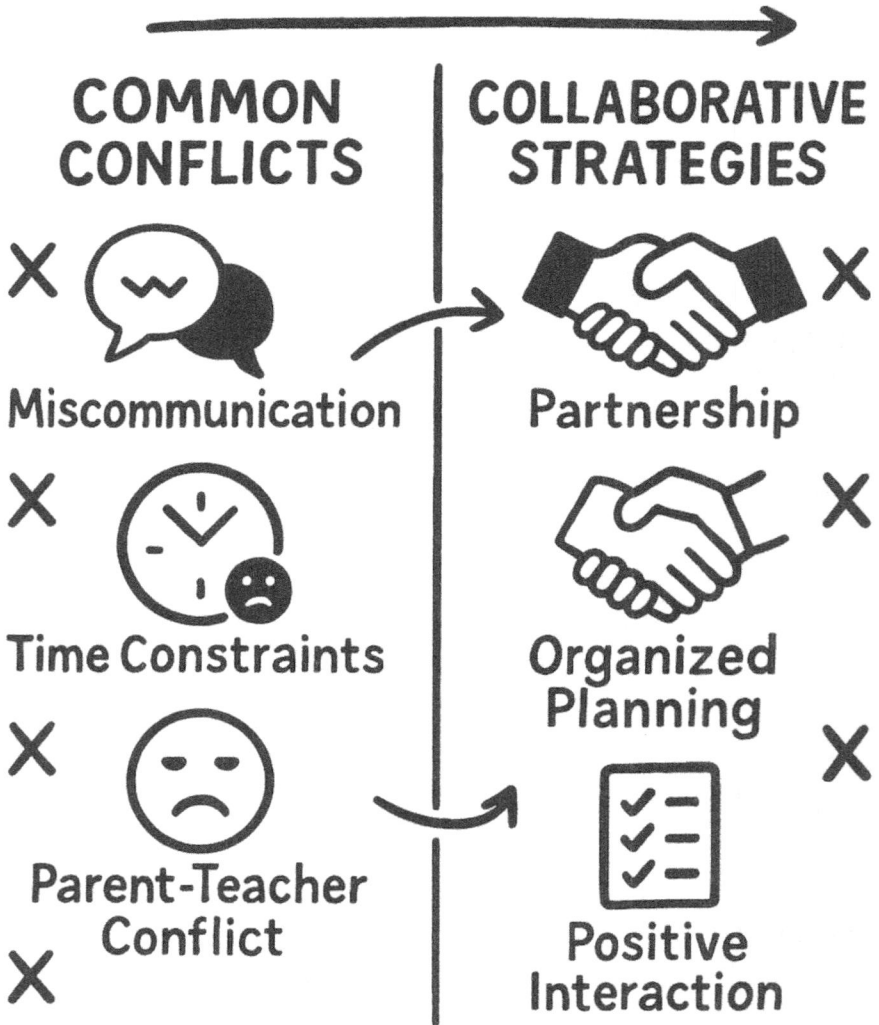

COMMON CONFLICTS	COLLABORATIVE STRATEGIES
Miscommunication	Partnership
Time Constraints	Organized Planning
Parent-Teacher Conflict	Positive Interaction

CHAPTER 8

From Conflict to Collaboration — Making the IEP Team Work for Your Child

I f you've ever walked into an IEP meeting feeling like you had to fight for every service or support, you're not alone.

Too often, what's supposed to be a team effort can feel more like a standoff.

Tensions rise. Trust breaks down. And somewhere in the middle, your child's needs get *lost*.

IEP meetings should feel like a team huddle to help your child win the game. But too often, they feel like a showdown.

However, it doesn't have to stay that way. We can change things.

With the right tools and mindset, you can move from conflict to collaboration and create an IEP team that works *with* you, not against you.

Why Conflict Happens

Let's be honest: if you've had hard experiences with the IEP process, it's not because you're "too emotional" or "asking for too much."

Here are some common reasons things go off track:

- The team is overwhelmed, understaffed, rushed, or stretched too thin.
- You've been dismissed, ignored, or talked down to.
- Promises were made—but not followed through.
- The IEP doesn't reflect what you *know* your child truly needs.

All of these frustrations are valid.

But staying stuck in conflict, especially year after year, can wear you down and stall progress.

Instead, let's shift from frustration toward something more powerful: informed advocacy. It is this strategic shift that wins the game.

5 Strategies to Move from Conflict to Collaboration

1. Focus on the Child

When emotions run high, anchor the conversation in this one question:

"What does [child's name] need to succeed?"

QUICK TIP

Use observations, examples, or data, not just frustration. For example: "At home, we've seen [skill] improve when we use [strategy]. Could we try something similar in class?"

2. Come With a Plan

When you show up with priorities, possible solutions, and a calm tone, you're seen as prepared, not combative.

QUICK TIP

Use your responses from the 40 Questions in Chapter 4 or your Leverage in Learning Plan to guide your requests. Even one clear goal suggestion can shift the tone of the meeting.

3. Document Everything

DID YOU KNOW?
Keeping a simple log of your communications and concerns can protect your rights—and it's often the first thing reviewed during disputes or mediation.

Verbal requests are easy to forget—or ignore altogether.

Follow up in writing to keep a clear record of what was said and when.

QUICK TIP

Keep a simple "parent log" with:
- Dates
- Names of who you spoke with
- Summary of what was discussed

It doesn't need to be fancy… just consistent.

4. Acknowledge Their Role

Most IEP teams are trying to do their best, but they're tired, stretched thin, and often criticized.

A little recognition goes a long way.

QUICK TIP

Start the meeting with a small acknowledgment.
"Thanks for being here and for the work you've done.
I know we may see things differently, but I appreciate the effort."
Appreciation builds trust.
And trust builds progress.

5. Know When to Escalate – Respectfully

If you've tried to collaborate and things still aren't working, you have legal rights.

These include mediation, due process, and state complaints.

But use these tools as a last resort, not your first strategy.

> **COACHING PRO TIP**
>
> Before escalating, ask:
> "Is there someone else who can help us find a solution –
> maybe a district specialist or facilitator?"

This shows you're still trying to solve the problem *together*, not going to war.

Fix it: When the Team Still Won't Collaborate

If you've come in with data, a clear tone, and thoughtful input—and the team is still resistant—don't go silent.

Try this strategy:

"It seems like we're having trouble finding common ground. I'd like to formally request a facilitated IEP meeting or bring in a neutral party to help us move forward."

Then, follow up in writing. Ask the school to document their decisions and reasoning in the Prior Written Notice (PWN). This often prompts a shift toward collaboration—or creates a clear trail for further action if needed.

Collaboration Isn't Backing Down – It's Leading with Clarity

Being collaborative doesn't mean being quiet.

It means being:

- Prepared

- Focused

- Firm (but respectful)

When you show up this way, you invite the rest of the team to do the same.

And that's how you build an IEP process that actually serves your child.

Even small shifts in tone and preparation can transform a hard meeting into a productive one.

You don't have to play defense in these meetings. "

Walk in with a plan, lead with clarity, and build collaboration with a team that truly supports your child.

You've got more power than you think, and your child deserves a team that uses it well.

NEXT STEPS

How to Prepare Proactively
- Review your top 3 IEP priorities before the meeting
- Write out how you plan to introduce them clearly and calmly
- Prepare 1–2 data points or examples to support each point
- Bring printed copies of your parent statement or plan
- Jot down one sentence of appreciation to open the meeting with strength—not stress

COACH'S CORNER

Key Plays to Keep in Mind

- Anchor the conversation in your child's needs—not your frustration
- Be prepared with priorities and solutions—not just concerns
- Know your options if collaboration stalls—your rights don't end at "no"

What this chapter helped you do

This chapter equipped you with strategies to shift IEP meetings from being tense to being collaborative.

You now know how to lead with clarity, document with purpose, and advocate with strength—even when things get tough.

FROM QUESTIONNAIRE TO CUSTOMIZED IEP

1 PARENT COMPLETES A DETAILED QUESTIONNAIRE

2 LEVERAGE IN LEARNING CREATES AN A.I. CUSTOM-ZIZED IEP PLAN

3 CUSTOMIZED IEP PLAN IS DISCUSSED AND FINALIZED BY THE IEP TEAM.

LEVERAGE IN LEARNING

.

CHAPTER 9

What Success Looks Like — and How to Keep It Going

An IEP isn't the finish line.

It's the starting point.

Think of an IEP as the kickoff; your child's growth is the real game, and you're leading the charge.

The real measure of success isn't in how detailed the paperwork is; it's in how your child grows. And that growth can take many forms, most of them quiet, beautiful, and deeply personal.

What Success Might Look Like

QUICK TIP

Don't wait for huge milestones. Document and celebrate small victories—they're the stepping stones to larger growth.
Not every milestone is measured in test scores or grades.

Here's what meaningful progress might look like:

- Your child reads a book independently – for the first time.
- They ask for help instead of shutting down.
- They initiate a conversation with a peer.
- They finish a task without giving up.
- They come home proud of something they accomplished.

These moments may not show up on a progress report, but they matter deeply.

Remember: Success isn't perfection.
It's movement. Growth. Confidence. Joy.

How to Know It's Working

Use these guiding questions to check in:

- Is my child more confident, curious, or willing to try?

- Are they making steady progress on their IEP goals?

- Am I hearing regularly from the school, especially about what's going well?

- Are we adjusting supports when something's not working?

If you're seeing 'yes' to even a few of these, you're moving in the right direction!

Every 'yes' means you're winning small victories. Stack them up, and progress becomes unstoppable.

Keeping the Momentum Going

Progress doesn't happen by accident; it grows when families stay engaged.

Here's how to keep things moving forward:

Schedule Regular Check-Ins

Don't wait for the annual IEP review. You can request a check-in any time, especially if something feels off or a goal needs adjusting.

Celebrate Small Wins

Even a tiny breakthrough deserves recognition.

"You finished your reading log without reminders. Look at that *focus!*"

Celebration builds motivation. Motivation fuels progress.

Advocate with Data

Bring more than opinions to meetings. Bring:

- Work samples

- Notes from home

- Behavior logs

- Tutor or therapist feedback

Real examples help the team see what's working… and what's not.

Adjust When Needed

An IEP isn't written in stone. If your child isn't progressing, you don't have to wait until next year.

You can call a meeting at any time to revisit goals, services, or accommodations.

Fix it: When Progress Feels Too Slow

Sometimes, despite everyone's best efforts, progress feels minimal—or even inconsistent. That doesn't mean failure. It means it's time to reframe expectations.

Try this:

"Instead of looking for big changes every month, I'll look for one sign each week that my child is trying, engaging, or bouncing back after a hard moment."

Then, bring it into focus:

- Write it down.

- Share it at your next IEP meeting.

- Ask, "What adjustments can we make to support continued growth?"

Progress is often messy. The key is recognizing and responding to what's really happening—not what we hoped would happen.

You're the Navigator!

You're not just following a plan—you're coaching your child toward success, step by step.

An IEP is a living roadmap.

It will change. Your child will change.

And that's the point.

As a parent, you bring the long view.

You see the day-to-day. The breakthroughs. The backslides. The becoming.

Stay engaged. Keep showing up. Celebrate what matters.

You're not just supporting an IEP. You're helping shape a future.

And your child is already on their way.

> **NEXT STEPS**
>
> **What You Can Do This Week**
> - Identify **one area** where your child has made small but meaningful progress
> - Celebrate it—out loud, with joy, in front of others
> - Make a note to share that win at your next IEP meeting
> - Reflect: Is there a goal that needs adjusting, based on what you've observed at home?
>
> Even one small action builds momentum.

COACH'S CORNER

Key Plays to Keep in Mind

- Success is more than scores—it's confidence, effort, and growth
- Track and celebrate progress, even when it's slow or uneven
- Use what you see at home to help the IEP team make smarter decisions

What this chapter helped you do

This chapter helped you recognize what meaningful progress looks like, advocate effectively with what you see, and sustain your child's momentum as they move forward with their IEP. You now have the tools to stay grounded in the long view—and to celebrate the wins that truly matter.

THE A.I. ADVANTAGE

SPEED WITH PURPOSE
A.I. scans documents quickly.

PATTERN RECOGNITION
A.I. spots repeated concerns.

LANGUAGE REFINEMENT
A.I. suggests specific wording.

ACCESSIBLE KNOWLEDGE
A.I. explains terms clearly.

THE A.I. ADVANTAGE

CHAPTER 10

You've Got This —
And You're Not Alone

Y ou've made it to the end of this first half—and what a game you've played! You've learned the fundamentals, practiced new plays, and started thinking like a pro advocate. But remember, halftime isn't the end. It's the reset, the regroup, the moment to take a breath and get back on the field stronger.

You're not starting from scratch anymore. You now know how the IEP process works, who's on the team, and how to make your voice heard. You've explored your child's strengths, named your concerns, and even gathered new tools to help shape a plan that fits.

By now, you've likely realized that being a parent in the IEP process isn't about knowing every answer—it's about knowing how to ask the right questions, who to turn to, and how to stay grounded in what your child needs. That's the real superpower.

We've started hinting at how Artificial Intelligence (A.I.) can be a helpful partner in this journey. In the next section, we'll go deeper into how A.I. can save you time, clarify confusing jargon, and even suggest language for tough conversations. You won't need to be a tech expert—just a curious parent who's ready to explore new tools.

But before we head into that next phase, take a moment to recognize how far you've come.

✎ QUICK TIP

Revisit your notes from the earlier chapters. What surprised you? What gave you hope? What still feels unclear? Jot those thoughts down—you'll carry them into the next section.

✅ ACTION STEP

CELEBRATE PROGRESS

- Find one moment from the IEP process that you handled better than before—big or small.

- Write it down. Tape it to the fridge or share it with someone who's cheering you on.

- Say this out loud: *"I may not have all the answers, but I have a plan—and I believe in it."*

📖 What this chapter helped you do

Realize how much knowledge and power you've gained. ✅ Reflect on your progress and prepare for next-level strategies. ✅ Feel confident stepping into the A.I. section as a more informed, prepared, and empowered parent.

COACH'S CORNER

You've trained hard through the fundamentals. Now it's time to explore what's next. The second half of this book brings fresh tools, new strategies, and a modern way to take your advocacy to the next level.

You're not alone, and you've already proven you've got what it takes.

Let's head into the next chapter—and level up.

A.I. CAN BE A STAR PLAYER IN PLANNING FOR YOUR CHILD'S SUCCESS.

A.I. SAVES REVIEW TIME ON COMPLEX INFORMATION TO FOCUS ON DETAILS OF YOUR CHILD AND BOOST ACADEMIC OUTCOMES.

CHAPTER 11

The A.I. Advantage

You've already done some heavy lifting by making it this far. As your coach, I want to pause and acknowledge that. Understanding the IEP process takes grit—and you've shown up.

Now, let's talk about something that might feel a little unfamiliar at first but has the power to completely shift your experience as a parent advocate: **Artificial Intelligence**, or A.I.

I know, it might sound like something from a tech lab or a sci-fi movie. But I want you to think of it differently—with me by your side. A.I. isn't a replacement for you, and it's not meant to make the process impersonal. It's a **tool**—and when we use it with intention, it becomes part of your coaching team.

What Is the "A.I. Advantage"?

Here's how I frame it for the families I work with: A.I. is a behind-the-scenes teammate. It's not calling the plays, but it's handing you the best data before the meeting. And most importantly, it responds to **your leadership**.

You bring the insights. A.I. brings the speed, the memory, and the translation. Together, that's what we call the *A.I. Advantage*.

Let me walk you through how I help families use it:

- **Speed with Purpose**

  When we're reviewing an IEP, A.I. can scan it in seconds. Together, we can see where the vague goals are hiding or which services don't line up with your child's needs. What used to take hours of flipping pages and second-guessing—now becomes focused and efficient.

- **Pattern Recognition**

  I once helped a parent spot a repeated reduction in services—speech, specifically—across three IEPs. There was no data to justify it. A.I. picked up on that instantly, and it gave us exactly what we needed to ask the right questions. That's not magic—that's momentum.

- **Language Refinement**

  How often do I hear this: "I know what I want to say... I just don't know how to say it." That's where A.I. comes in. It helps you shape goals that are measurable, parent input that's polished, and emails that reflect your message without losing your tone.

- **Accessible Knowledge**

  IDEA regulations are dense. Even as a coach, I refer to them often. But A.I. helps break them down into plain language. You don't have to memorize the law to understand it. A.I. helps us translate it so you can make informed decisions with confidence.

TINY STORY

Coaching Jason to Clarity

Let me share a story. Jason is a dad of a middle schooler with autism. Smart guy. Deeply invested. But every time he opened his son's IEP, he'd sigh and say, "I don't even know what I'm looking at."

So, we sat together and walked through it—line by line. I brought in a basic A.I. assistant to help highlight vague phrases and unclear services. Jason watched as we rewrote just one goal to make it specific and trackable.

At the next meeting, Jason asked:

"Can we revise this goal to include how you'll track progress?"

The room got quiet—and then the team said yes.

Later he told me, "I didn't feel like I was arguing. I felt like I finally knew what I was talking about."

That's what this is about. Not sounding perfect. Just feeling prepared.

COACH'S CORNER

Why A.I. Matters to You

Here's what I want you to remember:

- **A.I. remembers what you don't have time to.** It spots trends, flags inconsistencies, and connects dots across time.
- **It's emotionally neutral.** You show up with passion— A.I. shows up with data. Together, you're powerful.
- **It saves you time.** While you're busy parenting, it's already sorting the information you need.

This is one of the smartest ways I've seen parents reclaim their time and reduce burnout—without losing control.

ACTION STEP

Sharpen Your IEP Eye

Let's try a coaching exercise together:

1. Grab your child's most recent IEP and choose one goal.

2. Read it out loud.

   Ask yourself:

   - Does it clearly state what your child will do?

- Is there a way to measure success?

- Would someone outside the school understand what it means?

3. If any of your answers are "not really," then you've found a great place to test a support strategy. Open a free A.I. writing tool and ask:
   "Can you help me rewrite this goal to make it clearer and more measurable?"

Compare the two versions. Which one gives you more clarity? Which one sounds more like you?

Remember, you don't need to be a tech expert. You just need to be open to new tools that can amplify your advocacy.

DID YOU KNOW?

In 1924, an early version of A.I. was introduced in education—it wasn't digital, but it was a machine that gave students instant feedback. It helped personalize learning, just like today's tools do. Nearly a hundred years later, we're still chasing the same goal: giving families better ways to support learning.

What this chapter helped you do

This chapter gave you a front-row look at the role A.I. can play in making your voice stronger, your time better spent, and your plans more precise.

You explored:

- Why A.I. is worth considering as an advocacy partner

- The ways A.I. enhances—not replaces—your leadership

- A simple coaching activity to evaluate and improve one IEP goal

As your coach, I want you to walk away knowing this:

You're still the lead. A.I. is just here to make sure you're never left guessing.

WHAT A.I. CAN BE FOR YOU.

- Find the Gaps
- Spot the Patterns
- Improve Your IEP

A.I. HELPS TRANSLATE

CHAPTER 12

What is A.I. and Why It Helps

A s your coach, let me clear something up right away: A.I. doesn't replace your judgment—it strengthens it. Think of A.I. as your behind-the-scenes strategist. You bring the heart and the knowledge of your child. A.I. brings the organization, speed, and clarity that help you lead the conversation with confidence.

WHAT IS A.I. AND WHY IT HELPS?

Artificial Intelligence (A.I.) might sound like something out of a science fiction film—but here's the real story: A.I. is simply a tool that learns from patterns and data to support human decision-making. In our case, it's a tool that helps you become an even stronger advocate for your child's education.

Let me walk you through a few of the ways I've seen parents benefit from using A.I.—especially when paired with the insight only you have as a parent.

- **It breaks things down.** IEPs can feel like information overload. A.I. can sift through long documents and highlight what actually matters—so you don't waste time chasing details that don't move the needle.

- **It helps translate.** Technical or legal language can get in the way. A.I. can rephrase complex content into plain English that makes sense to you—and that you can confidently use in meetings.

- **It saves time.** Got a question about your child's services at 11 p.m.? A.I. tools are available 24/7 to provide instant, thoughtful responses based on best practices and data.

- **It echoes your voice.** One of the most powerful things about A.I. is its ability to work with your own words and priorities. You stay in control—but you speak with added clarity and precision.

When I think of A.I. in this space, I think of a coaching headset. You're still on the field making the plays, but A.I. is in your ear, helping you stay focused, strategic, and informed.

TINY STORY

TINA TAKES CHARGE

Let me tell you about Tina. She's a single mom of two kids and, like many parents, found herself swimming in IEP paperwork late at night—Googling acronyms, rereading sections, and still feeling unsure.

During one of our coaching calls, I introduced her to a simple A.I. tool. Together, we used it to summarize the IEP and pull out the three most important questions to bring to her next team meeting.

When the day came, Tina didn't just attend—she *led*. She asked about missing progress data and suggested a clearer way to track her son's reading growth. Her team listened, changes were made, and Tina left the meeting feeling confident—not confused.

She didn't need a degree in special education or in A.I. She just needed someone to show her how to use the tools at her fingertips.

COACHING PRO TIP

WHAT TO LOOK FOR

Here's how to get started with A.I. in a low-stress, high-impact way:

- **Start small.** Try using A.I. to rewrite a single sentence or goal that feels unclear. You don't need to overhaul everything.
- **Use it as a mirror.** A.I. helps reflect your message—not replace it. Use it to test how your concerns might be heard by others.
- **Stay in the driver's seat.** A.I. might suggest a route, but *you* know the destination. Trust your instincts and use A.I. as support, not authority.

ACTION STEP: TRY THIS

Choose one paragraph in your child's IEP that makes your eyes glaze over. You know the one.

1. First, rewrite it in your own words—as if you were explaining it to a friend.

2. Next, open an A.I. writing tool (many are free!) and ask: *"Can you help me make this paragraph clearer for an IEP meeting?"*

3. Compare the results. What got better? What would you keep from your version?

This simple exercise is how we start building muscle. And remember: I'm coaching you to experiment, not to be perfect.

✏️ QUICK TIP

Feeling unsure about where to begin? That's okay. Start by understanding key terms that often show up in meetings. One powerful example is:

Least Restrictive Environment (LRE)

Your child has the right to learn alongside peers without disabilities to the *maximum extent appropriate*. Special classes or separate settings are only used when a general education classroom—even with supports—won't meet your child's needs.

Understanding concepts like this helps you ask sharper questions and spot vague language that needs fixing.

? DID YOU KNOW?

The world's first chatbot, ELIZA, was built way back in 1966! It wasn't advanced, but people were amazed that a machine could mimic conversation. Today's A.I. tools are far more powerful—but they still serve a similar purpose: listening, responding, and helping people feel heard.

📖 What this chapter helped you do

This chapter gave you a simple but powerful introduction to how A.I. can help you in IEP planning:

- You learned how A.I. supports—not replaces—your voice.

- You discovered four practical ways it can simplify your advocacy work.

- You tried a hands-on strategy to clarify real IEP content.

You don't need to master A.I. overnight. Just take one step at a time—and know that I'm right here, coaching you along the way.

WATCH OUT FOR THESE COMMON RED FLAGS

EXCEPTIONS

E - Excessively Vague Goals
X - Lack of Data (the 'X)
C - Complex Jargon
Extra Supports Not Addressed

COACH'S CORNER

Unclear language makes it hard to know if the IEP is working. Don't be afraid to speak up.

A.I. IEP ADVANTAGE

CHAPTER 13

What A.I.
Can Do for You

I magine stepping into a meeting with a coach at your side—one who's already reviewed every play, spotted the gaps, and helped you prepare your next move. That's what A.I. brings to your IEP advocacy game. It's not about replacing your voice—it's about amplifying it with sharp, strategic insight.

When your deep knowledge of your child meets A.I.'s ability to analyze data, patterns, and language instantly, something powerful happens: you become an even more prepared, confident, and focused advocate.

WHAT CAN A.I. ACTUALLY DO?

You may be thinking, "Alright, I understand A.I. can help—but how, exactly?"

As your coach, let's walk through how A.I. strengthens your advocacy muscle, step by step:

- **Find the Gaps:** A.I. can scan your child's IEP and instantly highlight what's missing—whether it's vague goals, unclear progress monitoring, or absent services.

- **Spot the Patterns:** Looking across evaluations, reports, and previous IEPs, A.I. sees the big picture. If your child's reading scores haven't changed in years, A.I. flags it—so you can ask why.

- **Improve the Language:** Many IEPs include statements like "improve behavior" or "gain skills"—but what does that mean? A.I. can refine this language so it's measurable and actionable: "Demonstrate use of coping strategies in 4 out of 5 opportunities during structured tasks."

- **Provide Strategy Libraries:** A.I. can suggest evidence-based interventions based on the goal area. If organization is a concern, it might suggest visual checklists, color-coded folders, or daily planner routines.

- **Clarify Legal Alignment:** Unsure if the IEP meets IDEA standards? A.I. can summarize relevant regulations, helping you ask sharper questions—and advocate with confidence.

- **Prep Powerful Questions:** Before your next meeting, A.I. can generate targeted questions to help you focus on what matters most.

TINY STORY

MARIA'S BREAKTHROUGH

Maria, a mom of a fifth grader with dyslexia, had seen four IEPs come and go with little change. Together, we loaded her son's most recent IEP into an A.I.-powered tool. The results were eye-opening.

A.I. flagged vague goals, missing baselines, and a lack of meaningful progress tracking. It even suggested stronger language and helpful services to request.

When Maria attended her next meeting, she wasn't just informed—she was *in charge*. "I finally felt like the expert," she told me. "The team listened, not because I was loud, but because I was clear."

That's the power of preparation.

COACH'S CORNER

THREE BIG WINS WITH A.I.
- **Clarity = Confidence:** When you clearly understand your child's needs and the gaps in the IEP, you walk into the room stronger.
- **Focus Your Energy:** A.I. narrows your attention to the most important issues—so your time is spent where it matters.
- **Spot Missed Opportunities:** Sometimes, the IEP skips over important support areas. A.I. helps you catch what others might miss.

ACTION STEP

TRY THIS

Pick one section from your child's IEP—maybe the goals, services, or present levels. Read it with these three coaching questions in mind:

- Is the language specific and easy to understand?

- Is it measurable—can I tell when my child succeeds?

- Do I understand how progress will be tracked?

If you answer "no" to any, try rewriting the section in your own words. Then, run it by a free A.I. tool to get suggestions for improvement. You might be surprised how much sharper your version becomes.

QUICK TIP

If a sentence in the IEP feels vague or sounds like fluff, it probably is. A.I. can decode the jargon and help you say, "This isn't clear—can we be more specific?"

DID YOU KNOW?

The earliest form of A.I. in education dates back to the 1920s—a mechanical teaching machine that gave students immediate feedback. A century later, today's A.I. can do everything from language analysis to pattern tracking… and yes, it still offers timely, empowering feedback for learners and parents alike.

What this chapter helped you do

In this chapter, you explored:

- Real ways A.I. can support your IEP review and planning

- A real-life example of transformation through A.I. insight

- An easy action step to test out A.I.'s capabilities

You now know that A.I. can be your assistant coach—helping you lead, prepare, and elevate your advocacy.

SHOW UP PREPARED FOR IEP MEETINGS

- REVIEW THE IEP IN ADVANCE

- BRING A LIST OF CONCERNS AND REQUESTS

- ENSURE THAT YOUR INPUT IS INCLUDED

SHOW UP PREPARED

CHAPTER 14

Red Flags in the IEP — And What to Say

S potting red flags in an IEP can feel like trying to find a needle in a haystack—especially when the language sounds polished and official. But here's the coaching truth: vague doesn't mean valid, and confusing doesn't mean correct.

Your job isn't to write the IEP. It's to *understand it well enough* to ask clear, smart questions. That's where A.I. can help you see patterns—and where this chapter helps you find your voice.

This chapter is not about turning you into a lawyer or a teacher. It's about giving you the confidence and the tools to raise your hand and say, "Wait—can we make this clearer for my child?" Remember, your questions are not obstacles. They are invitations for collaboration.

TINY STORY

When Rosa first reviewed her son Mateo's IEP, she felt overwhelmed by the official-sounding language. One phrase stood out: "Will improve math skills as needed." Rosa wasn't sure what that meant, but it didn't sit right. At their next meeting, she spoke up: "Can we clarify what specific math skills Mateo is working on—and how we'll measure improvement?" That simple question opened up a productive discussion, and the team rewrote the goal to include concrete skills and a clear progress measure. Rosa says, "I used to stay quiet because I didn't want to seem pushy. Now I know asking for clarity helps everyone."

PARENT VOICE

"There was a time when I'd sign whatever was put in front of me because I didn't want to make waves. But I've learned that 'official language' doesn't always mean it's right for my child. Now, if something sounds vague or confusing, I pause and say, 'Can we make this clearer for my child?' That phrase has become my go-to. It shows I'm here to work with the team—but I'm also here to make sure nothing gets lost in fancy words."

COMMON RED FLAGS TO WATCH FOR

Let's walk through a few IEP phrases that might look fine on the surface—but raise red flags beneath. As your coach, I'll break down what they might mean, why they matter, and what you can say in response.

RED FLAG
"Will improve reading skills."

CAUTION

What's wrong: It's too vague. What kind of reading? How will we know it's improved? Are we talking about decoding, fluency, or comprehension?

TRY ASKING

"Can we make this more specific? What kind of reading skills are we targeting—and how will success be measured?"

COACH'S CORNER

Vague goals like this are common. A.I. tools can help rewrite this into something specific like: "Given a grade-level text, the student will identify the main idea in 4 out of 5 opportunities across two settings."

RED FLAG
"As needed."

CAUTION

What's wrong: This leaves too much room for interpretation. When is support "needed"? Who decides?

TRY ASKING

"Can we clarify what situations trigger this support? Can you give a classroom example?"

COACH'S CORNER

"As needed" can sometimes result in support being inconsistently provided. Ask for specific examples or data indicators that prompt support.

RED FLAG
No baseline data.

CAUTION

What's wrong: If the IEP doesn't say where your child is starting from, how will we track growth?

TRY ASKING

"Can we include current data or examples of work that show where my child is right now?"

COACH'S CORNER

Baselines are like starting lines. Without them, there's no way to track meaningful progress.

RED FLAG
Accommodations without context.

CAUTION

What's wrong: Listing supports without when/why/how they're used often leads to underuse.

TRY ASKING

"When exactly will this be used? How will staff know when to apply it?"

COACH'S CORNER

Accommodations should be tools, not decorations. Context ensures they are implemented with purpose.

RED FLAG
No mention of social-emotional needs.

CAUTION

What's wrong: If your child struggles socially or emotionally, these should be addressed alongside academics.

TRY ASKING

"Can we include supports or goals for emotional regulation or social interaction?"

COACH'S CORNER

Behavior is often communication. If your child's challenges go beyond academics, their IEP should too.

ACTION STEP

REVIEW WITH A FRESH EYE

Choose one section of your child's IEP (like Present Levels, Goals, or Accommodations) and read through it using this red flag lens. Highlight phrases that are vague, passive, or hard to measure.

Ask yourself:

- Can I tell what my child is supposed to do?
- Can I measure it?
- Is there enough information to track progress?

If not, this is your cue to ask for a revision—not because you're being difficult, but because clarity serves your child best.

✓ ACTION STEP

USE A.I. AS A FLAG FINDER

You don't need to catch every red flag by yourself. A.I. tools can quickly review your IEP and highlight language that's vague or inconsistent.

Try this:

Copy a section of your child's IEP into a tool and ask, "Can you help me identify any vague language or missing components?"

Then compare that with your own read. The insights might surprise you—and boost your confidence.

COACHING PRO TIP

YOU'RE NOT BEING DIFFICULT
Parents often worry about being labeled as "that parent." Let me be clear:
Asking for clarity isn't being difficult. It's being *diligent*. You are your child's best advocate. And advocates ask questions, seek clarity, and push for better.

Every red flag you raise is a step closer to a stronger, more effective IEP.

? DID YOU KNOW?
The U.S. Department of Education has emphasized that IEPs must include measurable goals and services based on documented needs. If a goal or accommodation is unclear, it may not meet IDEA requirements.

You are not just asking to ask—you are ensuring your child receives what the law promises.

What this chapter helped you do

Spot vague or ineffective language in your child's IEP

- Ask focused, respectful questions that lead to improvement

- Use A.I. to support your review and decision-making

This is about clarity, not conflict. It's about stepping into the room with tools, a voice, and a plan.

WHEN TO CALL IN
EXPERTS

GET HELP IF YOU:

- Need legal support for an IEP dispute
- Want an advocate at an IEP meeting
- Feel uncertain about shaping your child's IEP
- Get overwhelmed using A.I. tools

COACH'S TIP
An expert can provide tailored advice and peace of mind.

CHAPTER 15

Phrases to Use at the IEP Table

One of the most powerful tools in any IEP meeting isn't an app or a form—it's the words you choose. Having the right language in the moment can turn confusion into clarity, hesitation into action, and conflict into collaboration.

As your coach, my job is to help you walk into those meetings ready—not just with questions in mind, but with specific phrases that communicate your role as a leader in your child's education.

This chapter isn't about memorizing a script. It's about understanding how to guide the conversation toward what matters most: clarity, inclusion, and real progress for your child.

FOCUSED ADVOCACY: KEY PHRASES TO KEEP HANDY

Let's walk through several situations you might face at the IEP table, along with phrases that can help.

Clarifying Goals and Services

When the team presents goals that feel unclear or general, don't hesitate to slow things down and ask:

- "Can we review the specific skills my child is expected to develop with this goal?"

- "How will we know when this goal is achieved?"

- "Can you show me the baseline data this goal is based on?"

> ## COACHING PRO TIP
>
> If a goal sounds vague, it probably is. Be the voice that asks for clarity—your child benefits when everyone knows exactly what the target is. Instead of just listening passively, aim to walk out of the meeting knowing exactly what success looks like.

Ensure That Your Input Is Included — Why It Matters and How to Say It

As your coach, I want to make sure you never leave the IEP table wondering if your voice was truly part of the plan. Too often, parents share valuable insights and requests during a meeting—only to discover later that those points didn't make it into the final IEP document.

Your perspective isn't just helpful; it's legally protected. IDEA requires parent participation as part of the IEP team, and that includes having your concerns and recommendations reflected in writing.

Here's why this matters: If it's not in writing, it's not enforceable. Good intentions and verbal agreements won't hold up if questions arise later about what was decided.

How to Ensure Your Input Is Documented:

- At the end of the meeting, say: "Before we finish, can we review the notes or meeting summary together to make sure my input is reflected?"

- Ask directly: "Is there a section of the IEP where parent concerns and priorities are recorded? Can we add my points there?"

- If a team member says your input will be considered but doesn't write it down, respond with: "I appreciate that. For clarity, can we include my suggestion in the parent input section so it's part of the official record?"

- Before signing anything, review the draft IEP carefully. Don't hesitate to ask for edits that reflect exactly what you shared.

COACHING PRO TIP

Treat the IEP document like a team playbook. If your plays—your ideas, requests, and priorities—aren't written down, the team can't execute them. Having your input in writing helps you advocate with strength and clarity. It ensures that the plan reflects not just what the school recommends, but also what you know about your child as their parent and primary coach.

Addressing Service Gaps or Concerns

Even with a carefully written IEP, gaps in services or support can—and do—happen. Your role as a parent advocate includes watching for these gaps and knowing how to respond when they show up.

TINY STORY

When Monica's son Eli was supposed to start receiving occupational therapy twice a week, she trusted that things were on track. But three weeks in, Eli casually mentioned, "We don't go to that room anymore." Monica checked her notes and realized no one had updated her. She emailed the teacher using one of her go-to phrases: "According to the IEP, my child should receive occupational therapy twice per week. Can we review how that's being provided?" A quick team meeting followed, and the missed sessions were made up. Monica says, "It felt empowering to speak up early—before months had passed."

> **PARENT VOICE**
>
> "As a parent, I've learned not to assume the system is running smoothly. I don't want to wait until a report card shows a problem. Now, I check in regularly and ask, 'Can we look at the service log together to make sure everything lines up with the IEP?' That question has opened up helpful conversations—and kept my child from falling through the cracks."

Here's how you can approach it:

- **Spot the Gaps Early:** Pay attention to patterns. For example, if your child was supposed to have speech therapy twice a week but mentions they only go once, that's a potential gap. Keep track of what's promised in the IEP versus what's actually happening.

- **Document Your Concerns:** Write down specific examples of missed services, lack of progress, or any concerns. This gives you clarity and shows the school team that your feedback is based on facts, not feelings.

- **Raise Concerns Promptly and Clearly:** Here are specific phrases that can help:

  - "According to the IEP, my child should receive _____. Can we review how that's being provided?"

  - "I've noticed [service] hasn't been happening as scheduled. Can you help me understand why?"

  - "Can we look at the service log together to make sure everything lines up with the IEP?"

- ◆ "What is the plan to make up missed services?"

- ◆ "Who is responsible for ensuring [service] happens as outlined in the IEP?"

- **Request an IEP Team Meeting if Needed:** If the gap isn't resolved through informal conversations, ask in writing for an IEP team meeting:

  - ◆ "I'd like to request an IEP team meeting to address concerns about service delivery."

  - ◆ "Can we schedule a meeting to review service implementation and progress?"

- **Know Your Rights:** Under IDEA, schools must provide the services agreed upon in the IEP. If gaps persist:

  - ◆ "I'd like to request mediation if we're unable to resolve this."

  - ◆ "If this isn't resolved through the IEP team, what is the formal complaint process?"

COACH'S CORNER

You are not being "difficult" by speaking up. You are being an effective advocate. Addressing gaps isn't about creating conflict—it's about ensuring your child receives what they need to succeed.

Quick Action Checklist:

- Compare IEP services to what's happening in real life.

- Keep written notes with dates and examples.

- Bring concerns to the team's attention early using clear, specific phrases.

- Follow up with written communication if needed.

- Know when to request a formal meeting.

Advocating for Review and Data

Keeping the team accountable is easier when review dates and progress monitoring are clear from the start:

- "When will we review this plan again? Can we schedule a check-in sooner if needed?"

- "What kind of progress monitoring will be shared with me, and how often?"

- "Can you show me the data the team is using to make decisions about services?"

QUICK TIP

Progress isn't just an end-of-year conversation. Keep things moving with check-ins.

Navigating Disagreements or Uncertainty

Even in difficult moments, staying calm and prepared makes all the difference:

- "I'd like to pause and review this section again before we finalize it."

- "Can we look at IDEA requirements together to make sure we're aligned?"

- "I'd like to request mediation if we're unable to agree today."

COACH'S CORNER

Calm doesn't mean passive. Speaking up doesn't make you difficult—it makes you diligent.

Build Your Personal Phrase List

ACTION STEP

Here's a coaching tip that's served many parents well: Before your next IEP meeting, write down the top five phrases from this chapter that feel most relevant to your child's situation. Keep them in your notes app, print them out, or jot them on an index card. That way, when emotions run high, you have your advocacy playbook ready.

DID YOU KNOW?

Most school teams respect parents more when they show up with clear questions and specific language. It shows preparation, not confrontation.

What this chapter helped you do

- Speak up with confidence using parent-friendly, professional phrases.

- Know how to ask for clarity, data, and documentation at each step.

- Maintain control and focus, even in moments of disagreement.

KEEP PUSHING FORWARD

 • CONSISTENT EFFORT

 • LEARNING

 • ADVOCACY

PROGRESS TAKES TIME—DON'T GIVE UP

CHAPTER 16

Knowing When to Ask for Help — How A.I. and Experts Can Support You

A s a parent, there's power in knowing when to ask for support. Even with the right questions and phrases in hand, there may be times when you need extra help reviewing your child's IEP, understanding a school's response, or finding the right wording for a request.

This is where combining your own advocacy skills with A.I. tools and expert support can make a real difference.

WHY ASKING FOR HELP IS SMART ADVOCACY

You are your child's best expert—but you don't have to do it all alone. Many experienced parents and education advocates use both technology and human experts as part of their IEP team.

When to Consider Extra Support:

- When you're preparing for a major IEP meeting (like an eligibility review or transition plan).

- When you notice repeated service gaps that aren't being resolved.

- When a proposed IEP feels too vague, overwhelming, or confusing.

- When you want a second set of eyes to review documents for clarity and compliance.

HOW A.I. TOOLS CAN HELP YOU STAY ORGANIZED

A.I. isn't here to replace your voice—it's here to amplify it.

Ways to Use A.I. Support:

- Reviewing draft IEP language for clarity and specificity.

- Creating checklists for follow-up after meetings.

- Drafting sample emails and letters using coaching-friendly language.

- Organizing notes from multiple IEP meetings into clear summaries.

Coaching Reminder:

Use A.I. as your assistant, not your replacement. A tool can highlight patterns and spot unclear language—but it's your insight and experience that make decisions meaningful.

WHEN TO REACH OUT TO HUMAN EXPERTS

Sometimes, a real conversation with an advocate, lawyer, or education specialist is what's needed. This doesn't mean things have gone wrong—it means you're taking the next smart step in advocating for your child. Reaching out doesn't replace your voice—it strengthens it.

Here's how to know when expert help may be useful:

- You've reviewed your child's IEP using this book and A.I. tools, but certain sections still feel unclear or incomplete.

- You're preparing for a particularly complex IEP meeting— such as transition planning, eligibility reevaluation, or dispute resolution.

- You want help ensuring that your parent input and concerns are reflected accurately in the official record.

- You're facing persistent service gaps or disagreements that haven't been resolved through team meetings.

If you've walked through the strategies in this book and feel like having professional support would make things clearer or easier, that's a natural part of the process. Some parents prefer to handle things on their own using the tools and checklists provided here. Others decide that hiring an expert—like a special education advocate or consultant—gives them added peace of mind.

If that feels right for you, I'm here to help. My goal isn't to replace your voice but to amplify it. The coaching style and strategies you've read throughout this guide are exactly what I bring into consultations with families. Whether it's reviewing a draft IEP, helping you prepare for a key meeting, or supporting you through a particularly challenging situation, having an extra layer of expertise can make a difference.

Where to Find Expert Support:

- Your state's Parent Training and Information (PTI) center.

- Special education advocacy organizations.

- Education attorneys or advocates specializing in IDEA.

- Your local school district's special education parent liaison.

KEY PHRASES WHEN ASKING FOR HELP

If you're reaching out to an expert, advocate, or even your child's teacher, here are some starter phrases to keep things focused and positive:

- "I'd like a second set of eyes on this IEP draft—can you help me review it?"

- "Can you help me clarify the language in this section to make sure it reflects my child's needs?"

- "I'm feeling unsure about the next steps—what do you recommend?"

- "What resources or tools do you suggest for parents reviewing an IEP?"

TINY STORY

After years of handling her daughter Layla's IEPs on her own, Denise finally reached out to a local advocacy center. "I was nervous at first," she says, "but once I had someone reviewing things with me, I saw details I'd been missing. It didn't make me less of a parent—it made me stronger."

PARENT VOICE

"I used to think I had to do it all by myself. Now I know that the strongest advocates know when to call in backup. Whether it's A.I., an advocate, or another parent, support makes the process feel less overwhelming—and more successful."

COACH'S CORNER

Smart parents know: strong advocacy isn't about going it alone. It's about building a team—using your voice, technology, and human experts together. Don't wait until you feel stuck to ask for help. Make it part of your routine.

✅ ACTION STEP

Make a list right now:

- Who would you call if you needed help reviewing an IEP?
- What A.I. tools have been most helpful to you?
- What support would make this easier next time?

Write it down and keep it handy. That way, you're ready—before things get overwhelming.

📖 What this chapter helped you do

- Recognize when extra support is needed.
- Use A.I. tools as part of your IEP review and advocacy plan.
- Build your expert team and know how to ask for help in a focused, clear way.

NEED HELP? YOU'RE NOT ALONE.

Leverage in Learning is here to walk with you, not ahead of you.

Visit www. leveragerinlearning.com or email me directly at A.I.IEP.Advantage@leveragein learning.com.

CHAPTER 17

Pulling It All Together — Your IEP Advantage in Action

By now, you've walked through the full playbook: from spotting red flags, to asking clear questions, to knowing when and how to get extra support. This final chapter is about taking what you've learned and putting it into steady, confident action.

WHAT ADVOCACY LOOKS LIKE OVER TIME

IEP advocacy isn't a one-time event. It's an ongoing process of checking in, reviewing, clarifying, and collaborating. Advocacy shifts and grows as your child grows. What you focus on in kindergarten might look different from what you focus on in middle school or during transition planning for adulthood. It's not just about reacting to problems—it's about proactively shaping your child's education plan over time.

Think of advocacy as a cycle: observe, reflect, ask, clarify, adjust, repeat. That rhythm becomes part of how you stay involved—not just when something urgent happens, but as part of your family's routine.

Signs You're Building Long-Term Advocacy Habits:

- You regularly review your child's IEP with fresh eyes each school year.

- You feel comfortable asking clarifying questions during meetings without hesitation.

- You've created a personal list of go-to phrases and checklists that work for your family.

- You balance using A.I. tools with trusting your own insights.

- You know when to reach out for help—and have trusted contacts ready.

MAINTAINING YOUR IEP ADVOCACY ROUTINE

Staying consistent with IEP advocacy doesn't have to feel overwhelming. The key is to build simple habits into your regular schedule—just like checking in on bills, doctor appointments, or family activities. It's about creating a steady rhythm that keeps you informed and prepared without waiting for a crisis.

Here's what I recommend as a basic routine:

- Set a calendar reminder to review your child's IEP document once each semester. Treat this as a non-negotiable check-up.

- Schedule regular check-ins with teachers and service providers—ideally once per quarter. These don't always have to be formal meetings; sometimes a brief email or phone call can keep things on track.

- Use A.I. or personal checklists to spot gaps, unclear language, or outdated information. A quick review every few months can save you from bigger issues later.

- Keep notes and progress samples organized in a folder or digital file. Having your own records makes it easier to prepare for meetings and spot trends.

- Review service logs or session notes if your district provides them. Don't assume services are happening exactly as written—verify and check.

- Refresh your list of go-to advocacy phrases and questions before each major IEP meeting. This helps you walk in feeling confident and ready.

- Make advocacy a shared routine if possible—engage another family member or trusted friend to review documents or attend meetings with you for support.

When these steps become part of your yearly and monthly rhythm, IEP advocacy feels less like a scramble and more like a steady part of parenting.

TINY STORY

After following these steps for two years, Michael shared: "At first I felt lost and overwhelmed. Now I have a routine. I check my son's IEP like I check my bank account—regularly and without fear. It's part of how I stay involved."

PARENT VOICE

"I realized advocacy isn't just about showing up to meetings. It's about building habits that keep me in the loop all year. Once I made it part of my routine, it felt less stressful—and more like just another part of parenting."

GIVING YOURSELF CREDIT

Advocacy can be invisible work. No one sees all the emails you send, the notes you keep, or the quiet moments where you review paperwork late at night.

I want you to pause here and acknowledge that effort. You've done the work to show up for your child in a powerful, informed way. Whether you choose to handle things independently, use A.I. tools, or bring in expert help—you've built a foundation that will serve your child well.

COACH'S CORNER

If you remember one thing from this guide, let it be this: Your voice matters.

You don't have to know everything. You don't have to handle it all perfectly. You just have to show up, ask clear questions, and keep the process moving forward.

ACTION STEP

Make a written commitment to yourself:

- I will review my child's IEP every semester.

- I will ask questions whenever something feels unclear.

- I will trust my instincts and seek support when needed.

Sign it, date it, and keep it in your advocacy folder. This isn't just paperwork. It's your plan to stay engaged and empowered.

What this chapter helped you do

- Maintain long-term advocacy habits.

- Balance independence, A.I. tools, and expert help.

- Stay confident, consistent, and ready to support your child's growth.

CHAPTER 18

Final Thoughts —
Your Role, Your Power,
Your Plan

As we close this guide, I want to leave you with one simple truth: You are already enough. By choosing to read, learn, and show up for your child's IEP journey, you've taken the most important step any parent can take. This book wasn't about turning you into a lawyer or a teacher. It was about showing you how to use your natural role as a parent—and combining it with modern tools like A.I. and expert advice—to create better outcomes for your child.

YOU'VE GOT THIS—AND YOU'RE NOT ALONE

You've reached the end of this guide, but you're really just at the beginning of a new chapter in your advocacy journey.

While you've crossed the finish line of this book, advocacy is a lifelong game… but now, you're ready for it!

If you've read this book, you now have something many parents don't: a roadmap, a language, and a sense of direction. You know what strong IEPs look like. You understand your rights. You've reflected on your child's strengths and challenges.

And you've seen how artificial intelligence, when paired with your insight, can become a powerful support tool at home and in meetings.

WHAT I HOPE YOU TAKE WITH YOU

- You don't have to know everything to advocate effectively.

- Clear, consistent questions are often more powerful than having all the answers.

- Technology can support you, but your insight as a parent is irreplaceable.

- Advocacy isn't about conflict—it's about clarity, collaboration, and action.

REMEMBER THE BALANCE

You now have:

- The confidence to ask focused, respectful questions.

- Tools like A.I. checklists and red-flag finders.

- Options for expert support when you want extra help.

Some parents will use this book and feel fully equipped to handle things on their own. Others may decide that partnering with an advocate or consultant (like myself) adds peace of mind. Both choices are valid—and the decision is always yours.

NEED HELP? YOU'RE NOT ALONE.

If at any point this feels overwhelming—if the process starts to blur, or your confidence slips—please know this: you don't have to do it by yourself.

Leverage in Learning was built to walk with you, not ahead of you. Whether you need a second opinion, a custom plan, or just someone to say "you're not crazy for feeling this way," we're here.

Visit www.leverageinlearning.com or email me directly at A.I.IEP. Advantage@leverageinlearning.com.

YOUR PERSONAL ADVOCACY PLAN

Before you close this book, take a moment to write out your own personal IEP advocacy plan. Here's a sample template to get you started:

My IEP Advocacy Plan

- I will review my child's IEP every _____.

- I will use A.I. tools or checklists to help me _____.

- I will ask for help when _____.

- My go-to expert contacts are _____.

FINAL WORD OF ENCOURAGEMENT

You are your child's most consistent advocate, guide, and believer.

You don't have to be perfect. You just have to keep showing up.

This book was written to equip you—not just with information, but with belief.

Belief in your ability to navigate this system. Belief that your child deserves more than the minimum. And belief that when families are informed and empowered, things change.

You're leading this charge, and you're more prepared than ever. Keep going, keep learning, stay in the game, and trust that every day, as long as you make progress, you are winning.

FINAL COACH'S CORNER

You've already done the hardest part: deciding to take this journey for your child. Trust yourself. Stay steady. And remember—you're never alone in this work. Thank you for letting me be part of your IEP journey.
— Dr. Al C. Jones, Jr.
Leverage in Learning

Glossary of Terms

A.I. (Artificial Intelligence) — Technology that simulates human thinking to analyze, predict, and assist with tasks like reviewing IEP language, spotting patterns, or suggesting clearer wording.

Accommodations — Adjustments in teaching methods, materials, or testing that allow a student with disabilities to access curriculum and demonstrate learning without changing content.

Advocacy — The process of actively supporting and standing up for your child's rights, services, and educational needs through informed participation.

Baseline Data — Information about a student's current skill levels or performance used as a starting point to measure progress on IEP goals.

Clarification Questions — Targeted questions asked during an IEP meeting to make language clearer or confirm understanding.

Collaboration — Working together with teachers, service providers, and other IEP team members to create the best plan for a student.

Complex Jargon — Overly complicated or technical language in IEP documents that may confuse parents or make expectations unclear.

Data Monitoring — The ongoing collection and review of information to track student progress toward IEP goals.

Determination Meeting — A meeting where the IEP team reviews data to decide whether a student continues to qualify for services under IDEA.

Due Process — A formal legal procedure available when disagreements arise about a student's IEP services or placement.

Exceptions (Red Flags) — Common warning signs in IEPs such as excessively vague goals, lack of baseline data, complex jargon, or missing supports.

Extended Services — Services provided beyond the regular school day or year to help meet a student's IEP goals.

Goals (Measurable IEP Goals) — Specific, detailed statements in the IEP that describe what a student is expected to learn or achieve within a set timeframe.

IEP (Individualized Education Program) — A legally binding document developed for students with disabilities outlining services, accommodations, and educational goals.

IEP Team — The group of individuals responsible for developing, reviewing, and revising a student's IEP. It includes parents, teachers, specialists, and sometimes the student.

Parent Input Section — A part of the IEP document where parent concerns, observations, and priorities are recorded.

Progress Monitoring — Regular checks on how well a student is meeting IEP goals using data collection and assessment tools.

Red Flags — Phrases or patterns in IEP documents that signal a need for review, clarification, or revision to better support the student.

Service Gaps — Missing or inconsistently provided services that are required by a student's IEP.

Specificity — The level of detail and clarity used in writing IEP goals, services, and accommodations to ensure understanding and accountability.

State Parent Training and Information Center (PTI) — Federally funded centers that provide resources, training, and support for families navigating special education.

Transition Planning — IEP planning that helps prepare students for life after school, including further education, employment, and independent living.

Vague Language — IEP wording that is unclear, non-specific, or open to interpretation, often requiring parent clarification.

Parent Coaching — Professional support offered to parents to strengthen advocacy skills, including help reviewing IEPs, preparing for meetings, and navigating special education systems.

Leverage in Learning — The author's parent support and coaching service designed to help families personalize and strengthen their IEP process using the strategies and tools from this book.

List of Abbreviations

A.I. — Artificial Intelligence

ADA — Americans with Disabilities Act

APE — Adapted Physical Education

ASD — Autism Spectrum Disorder

AT — Assistive Technology

BIP — Behavior Intervention Plan

DD — Developmental Delay

DMS — Differentiated Monitoring and Support

ED — Emotional Disturbance

ESY — Extended School Year

FAPE — Free Appropriate Public Education

FERPA — Family Educational Rights and Privacy Act

FIE — Full Individual Evaluation

FBA — Functional Behavior Assessment

IDEA — Individuals with Disabilities Education Act

IEE — Independent Educational Evaluation

IEP — Individualized Education Program

IFSP — Individualized Family Service Plan

IQ — Intelligence Quotient

LRE — Least Restrictive Environment

MTSS — Multi-Tiered System of Supports

OHI — Other Health Impairment

OSEP — Office of Special Education Programs

OT — Occupational Therapy

PT — Physical Therapy

PTI — Parent Training and Information Center

RTI — Response to Intervention

SEA — State Educational Agency

SEL — Social Emotional Learning

SPED — Special Education S

TA — Technical Assistance

www.ingramcontent.com/pod-product-compliance
Lightning Source LLC
Chambersburg PA
CBHW071436090426
42737CB00011B/1674